Polar Bears

Edited by Katie Gillespie

openlightbox.com

Go to www.openlightbox.com and enter this book's unique code.

ACCESS CODE

LBP55339

Lightbox is an all-inclusive digital solution for the teaching and learning of curriculum topics in an original, groundbreaking way. Lightbox is based on National Curriculum Standards.

STANDARD FEATURES OF LIGHTBOX

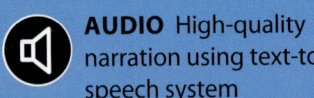 **AUDIO** High-quality narration using text-to-speech system

 ACTIVITIES Printable PDFs that can be emailed and graded

 SLIDESHOWS Pictorial overviews of key concepts

 VIDEOS Embedded high-definition video clips

 WEBLINKS Curated links to external, child-safe resources

 TRANSPARENCIES Step-by-step layering of maps, diagrams, charts, and timelines

 INTERACTIVE MAPS Interactive maps and aerial satellite imagery

 QUIZZES Ten multiple choice questions that are automatically graded and emailed for teacher assessment

 KEY WORDS Matching key concepts to their definitions

Copyright © 2017 Smartbook Media Inc. All rights reserved.

2 **Animals of North America**

Contents

Lightbox Access Code	2
Meet the Polar Bear	4
All about Polar Bears	6
Polar Bear History	8
Polar Bear Habitat	10
Polar Bear Features	12
What Do Polar Bears Eat?	14
Polar Bear Life Cycle	16
Encountering Polar Bears	18
Myths and Legends	20
Polar Bear Quiz	22
Key Words/Index	23
Log on to www.openlightbox.com	24

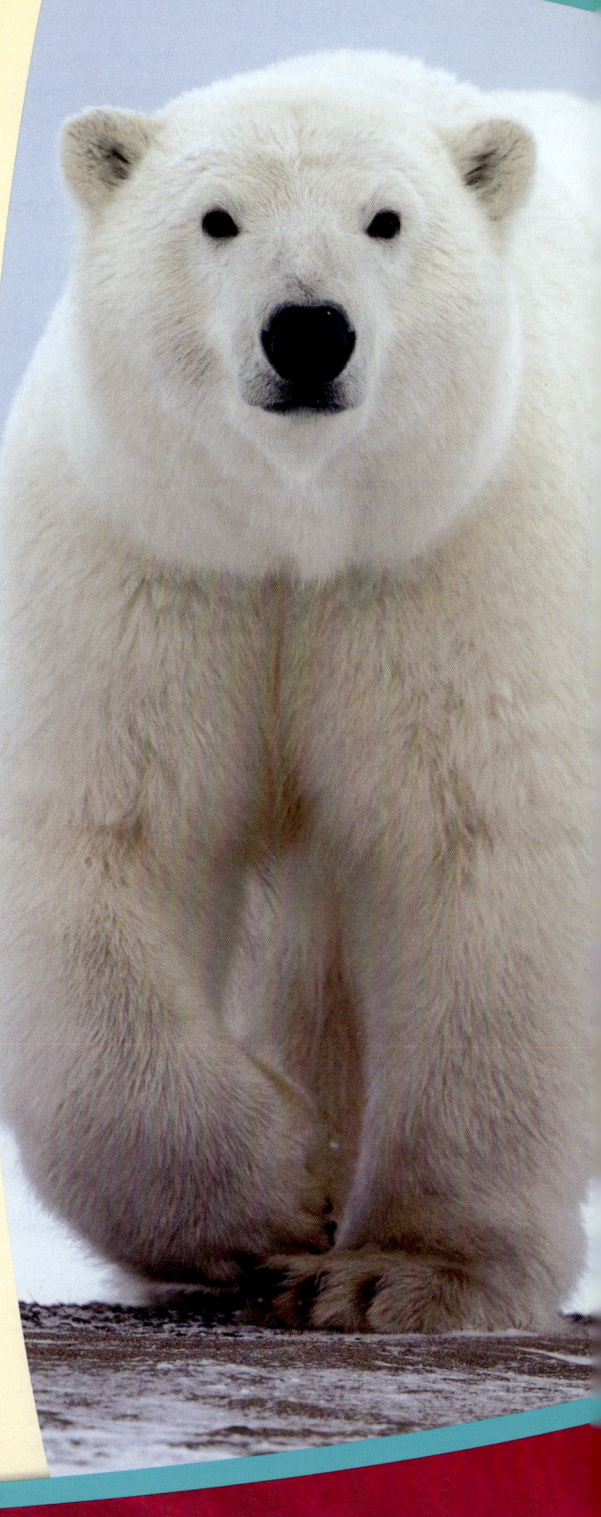

Polar Bears   3

Meet the Polar Bear

Polar bears are members of the bear family, *Ursidae*. They are the largest land **carnivores** in the world. An adult male polar bear can measure 8 to 10 feet (2.4 to 3 meters) long, and weigh between 750 and 1,600 pounds (340 and 720 kilograms).

Polar bears live in icy **Arctic** regions, where temperatures are below freezing most of the year. Their fur helps keep their bodies warm. This fur traps body heat while allowing sunlight to warm the bear's skin.

Polar bears are strong swimmers. Their paws act as natural paddles, pushing their bodies through the water. These Arctic bears spend more time in the water than any other bear.

> Polar bears can live 25 to 30 years in nature.

Animals of North America

Polar Bears 5

All about Polar Bears

North American polar bears live only in the northern part of the continent. This area is called the Arctic. Polar bears can be found in northern Alaska near the Beaufort and Chikchi seas, and as far south as James Bay in Canada. There are an estimated 20,000 to 25,000 polar bears spread throughout the Arctic.

While hunting on sea ice, adult male polar bears travel alone. Family groups consist of a mother and her cubs. Most of the ice melts in the summer and fall. During this time, polar bears stranded on land are sometimes seen together.

Polar Bear Skull

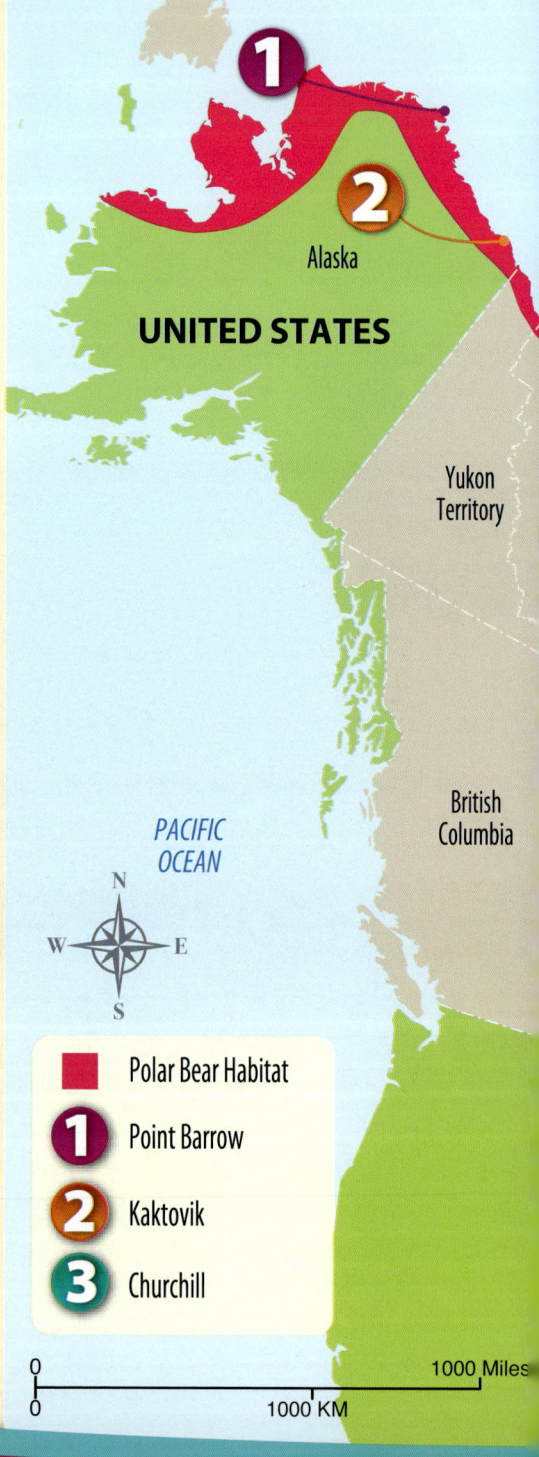

Polar Bear Habitat
1. Point Barrow
2. Kaktovik
3. Churchill

Animals of North America

Polar Bear History

Scientists believe that polar bears developed from the **ancestors** of brown bears roughly 500,000 to 750,000 years ago. The ancient bears became isolated by glaciers in ice-covered northern areas during the last **Ice Age**. These ancestors adapted to their new climate, learning to hunt seals and survive on the sea ice.

About 18,000 years ago, ice sheets covering the northern part of Earth stopped advancing southward. As the ice began to retreat north over several thousand years, polar bears and seals moved northward with it. Today, polar bears stay in the Arctic, where the ice is thick.

The cave bear is an extinct ancestor of the polar bear.

Polar Bear Timeline

120,000 Years Ago
Polar bears become carnivores, living almost exclusively on other marine life. Their ancient bear relatives were **omnivores**.

20,000 Years Ago
Polar bears develop teeth that are different from those of their brown bear ancestors. These teeth are ideal for biting through ice and hunting marine prey.

Today
The polar bear is listed as a threatened species by the U.S. government. Due to their slow reproduction rate, polar bear populations are still recovering after being hunted during the twentieth century.

500,000 to 750,000 Years Ago
A population of ancient brown bears become isolated in the north during the last Ice Age.

1774
Scientist Constantine J. Phipps gives the polar bear its scientific name, *Ursus maritimus*, meaning "sea bear."

1960s and 1970s
Polar bears are hunted to the brink of **extinction**.

1973
The International Agreement on the Conservation of Polar Bears is passed. This treaty protects polar bears by enforcing strict hunting regulations.

Polar Bears 9

Polar Bear Habitat

Polar bears spend most of their lives out on the sea ice and get their food from the sea. They have been found more than 100 miles (161 kilometers) from any land. Polar bears mainly stay close to their food source. As they hunt for seals, polar bears cover great distances. They travel farther in their lifetimes than any other land animals.

All polar bears have individual home ranges, or areas where they live, hunt, and sleep. A polar bear's home range can be large, sometimes up to 135,000 square miles (350,000 square km). The home range includes the polar bear's **migration** during summer. Some polars bears stay on the sea ice, moving north over the summer as it melts. Other polar bears come onto land and wait for winter, when the ice returns.

The **heaviest** polar bear ever recorded weighed **2,209** pounds. (803 kg)

Polar bears are only the size of a **kitten** at birth.

Nanuq means "polar bear" in the **Inuktitut** language.

Churchill, Canada, is called the **polar bear capital** of the world.

10 **Animals of North America**

Polar bears prefer to inhabit broken ice, as there are more seals surfacing to breathe in these areas.

Polar Bears 11

Polar Bear Features

Over thousands of years, polar bears have developed many special features that help them live and hunt in cold climates. These features, such as wide paws and thick, white fur, help polar bears flourish in the Arctic. They overheat very easily when the temperature is above freezing.

FUR
A polar bear's coat is made of two layers of fur. There is a dense, woolly undercoat and an outer layer of long, coarse hair, called guard hair. An oily coating on each guard hair helps **repel** water.

PAWS
Measuring up to 12 inches (30 centimeters) across, a polar bear's paws work like snowshoes to help the bear stay on top of the snow. Rough black pads help it walk on slippery ice.

CLAWS
Each of the five toes ends with a curved claw about 3 inches (7.5 cm) in length. The claws are used to catch prey. Their sharp tips also help prevent polar bears from slipping when traveling across ice.

Animals of North America

EYES

A polar bear's eyes can filter the Sun's glare off the snow. The bear's eyes have a clear inner eyelid that helps them see when underwater.

NOSE

Polar bears use their sense of smell to find food. They can smell prey 1 mile (1.6 km) away. Polar bears also use their sense of smell to find potential mates from tracks in the snow.

TEETH

A polar bear's teeth have sharp, jagged edges. This allows them to rip off bite-sized chunks of **blubber** and meat. The canine teeth are used for gripping and tearing the hides off prey.

Polar Bears 13

When on shore, polar bears will eat birds and other land animals.

14 Animals of North America

What Do Polar Bears Eat?

Polar bears are carnivores. Their most common meal is the ringed seal. Large adult polar bears can eat up to 90 pounds (41 kg) of food in one meal.

In addition to ringed seals, polar bears also hunt bearded seals. They are much larger, weighing 900 pounds (408 kg) or more. Polar bears occasionally eat other marine life, such as young walruses, beluga whales, or narwhals.

For polar bears trapped on shore when the sea ice melts, there is very little to eat. The bears will mainly try to find **carcasses** washed up on shore, birds' eggs, and small rodents such as lemmings and voles. They will also eat berries, kelp, and seaweed.

Polar bears will hunt ringed seals that are resting on the ice.

Polar Bear Life Cycle

One of the few times polar bears come together in groups is for mating season. This occurs on the sea ice in April or May. In October, the female polar bear enters a maternity den, where she will give birth and **hibernate** with her young. The baby polar bears, called cubs, are then born around December.

Birth

A mother polar bear will usually give birth to two cubs, but can have as many as four. Newborn cubs weigh about 21 to 25 ounces (600 to 700 grams). Their eyes are not yet open, and their pink skin is covered with white, fuzzy fur.

One to Five Months

By the end of their first month, the cubs can see and hear, and their teeth start to grow. Their mother leaves the den between late February and the middle of April. At this time, male cubs weigh 22 to 26 pounds (10 to 12 kg), while female cubs weigh slightly less.

Animals of North America

Adult

By the time they are about three years old, polar bears are considered adults. They can hunt for themselves. Female polar bears do not mate until they are four years of age or older. As long as they are not caring for a cub, they will mate every two to three years. Male polar bears will not begin mating until they are five or six years old.

Six Months to Three Years

By six months, the cubs are almost fully grown. They can keep up with their mother on hunts and swim easily. Food can be hard to find, however, so polar bear cubs stay with their mothers for two or three years. A mother polar bear will not **wean** her cubs until they are more than two years old.

Encountering Polar Bears

Polar bears are **solitary** creatures, but they do come into contact with humans occasionally. This most commonly occurs when polar bears are driven onto land by the melting ice. They may stray into human settlements to look for food.

All polar bears, including cubs, can be dangerous and unpredictable. A hungry polar bear will sometimes try to track a human. Never approach, feed, or disturb a polar bear.

Polar bears **greet** each other by **touching noses.**

A polar bear's skin is **black**, which helps it absorb **heat** from the Sun.

Polar bears have a layer of **fat** that can be **4.5 inches thick.** (11 cm)

 A polar bear can **hold** its **breath** underwater for up to **two minutes.**

18 Animals of North America

Polar bears have no natural predators, and no fear of humans.

Polar Bears 19

Myths and Legends

Polar bears are featured in many **Inuit** legends. These Aboriginal Peoples of the far north credit the polar bear with teaching man how to hunt seals. Polar bears are thought to be shape shifters, able to transform into humans and then back again into bears once they donned their bear coats.

The Inuit also believe that humans and animals have spirits that live on after death. Polar bears are said to have one of the most powerful animal spirits. After successfully hunting a polar bear, the Inuit would hold a great feast and perform a "polar bear dance" to show respect and thanks for the bear giving up its life to the hunter. It was believed that not showing respect would doom future hunts.

Inuit artists add humor and character to the polar bear by carving it in a dancing pose.

The Woman and Her Bear

Far to the north, there was a lonely old Inuit woman without a family. Though her neighbors were kind to her, she was lonesome. She prayed to the gods for a family of her own.

One day, she found an orphaned polar bear on the ice. She determined that the cub must be alone, so the old woman took the bear home and raised him as her son. The bear cub was soon beloved both by his mother and the village children. The children taught him to hunt and he became the best hunter in the village.

Soon, the old woman's neighbors became jealous and afraid of the bear. They envied how many fish and seals the bear caught for his mother and worried he was getting too big. They plotted to drive the polar bear from their village.

A boy from the village heard this talk and ran to tell the old woman. She begged her son to leave so that their neighbors would not harm him. The bear son heeded his mother, though he cried to leave her.

The old woman waited for her neighbors to forget about her bear son. Then, she set out onto the ice to find him. She traveled for many days and finally found him again. Out of love and kindness toward his mother, the polar bear came to visit her again and found her fish and seals to eat.

The villagers learned to understand that the bond between the old woman and her bear son was strong and true. To this day, they proudly tell the story of the unbroken love between mother and son.

Polar Bears

Polar Bear Quiz

1 In which region do polar bears live?

2 What color is a polar bear's skin?

3 What do polar bears mainly eat?

4 What is the polar bear's scientific name and what does it mean?

5 Where does a mother polar bear give birth to her cubs?

6 How long do polar bears stay with their mothers?

7 What is the Inuit word for polar bear?

8 How many polar bears live in nature?

9 How much can an adult male polar bear weigh?

10 What city is called the polar bear capital of the world?

Answers: 1. The Arctic 2. Black 3. Ringed seals 4. *Ursus maritimus*, "sea bear" 5. A maternity den 6. Two to three years 7. Nanuq 8. 20,000 to 25,000 9. Up to 1,600 pounds (725.7 kg) 10. Churchill

22 Animals of North America

Key Words

ancestors: relatives from the past

Arctic: the region of Earth around the North Pole where the climate is very cold

blubber: layers of fat in large sea mammals

carcasses: remains of a dead animal

carnivores: animals that eat mostly meat

extinction: a state of no longer living any place on Earth

hibernate: to sleep through the winter months

Ice Age: a time when most of Earth was covered with ice

Inuit: Aboriginal Peoples of northern Canada and parts of Alaska

migration: moving from one area to another

omnivores: animals that eat food of both plant and animal origin

repel: resist mixing or absorbing

solitary: existing alone

wean: to teach young how to live and find food on their own

Index

Arctic 4, 6, 8, 12, 22

cubs 6, 16, 17, 18, 21, 22

den 16, 22

eyes 13, 16

fur 4, 12, 16

hunting 6, 8, 9, 10, 12, 15, 17, 20, 21

ice 6, 8, 9, 10, 11, 12, 15, 16, 18, 21

Inuit 20, 21, 22

paws 4, 12

seals 8, 10, 11, 15, 20, 21, 22

teeth 9, 13, 16

➕ SUPPLEMENTARY RESOURCES

Click on the plus icon ➕ found in the bottom left corner of each spread to open additional teacher resources.

- Download and print the book's quizzes and activities
- Access curriculum correlations
- Explore additional web applications that enhance the Lightbox experience

LIGHTBOX DIGITAL TITLES
Packed full of integrated media

VIDEOS

INTERACTIVE MAPS
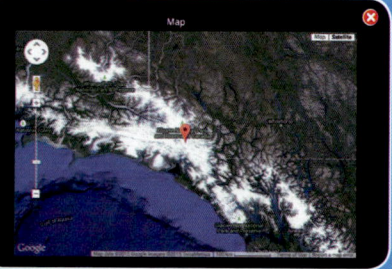

WEBLINKS

SLIDESHOWS

QUIZZES

OPTIMIZED FOR
✓ TABLETS
✓ WHITEBOARDS
✓ COMPUTERS
✓ AND MUCH MORE!

Published by Smartbook Media Inc.
350 5th Avenue, 59th Floor New York, NY 10118
Website: www.openlightbox.com

Copyright © 2017 Smartbook Media Inc.
All rights reserved. No part of this publication may be reproduced, stored in a retrieval system, or transmitted in any form or by any means, electronic, mechanical, photocopying, recording, or otherwise, without the prior written permission of the publisher.

Library of Congress Control Number: 2015953947

ISBN 978-1-5105-0818-7 (hardcover)
ISBN 978-1-5105-0820-0 (multi-user eBook)

Printed in the United States of America in Brainerd, Minnesota
1 2 3 4 5 6 7 8 9 0 19 18 17 16 15

112015
110515

Editor: Katie Gillespie
Designer: Mandy Christiansen

Every reasonable effort has been made to trace ownership and to obtain permission to reprint copyright material. The publisher would be pleased to have any errors or omissions brought to its attention so that they may be corrected in subsequent printings. The publisher acknowledges Getty Images, Corbis, and Alamy as its primary image suppliers for this title.

24 Animals of North America